THE STEAM TURBINE

THE STEAM TURBINE

THE REDE LECTURE
1911

BY

Sir CHARLES A. PARSONS, K.C.B.

Cambridge :
at the University Press
1911

CAMBRIDGE UNIVERSITY PRESS
Cambridge, New York, Melbourne, Madrid, Cape Town,
Singapore, São Paulo, Delhi, Tokyo, Mexico City

Cambridge University Press
The Edinburgh Building, Cambridge CB2 8RU, UK

Published in the United States of America by
Cambridge University Press, New York

www.cambridge.org
Information on this title: www.cambridge.org/9781107615090

© Cambridge University Press 1911

First published 1911
First paperback edition 2011

A catalogue record for this publication is available from the British Library

ISBN 978-1-107-61509-0 Paperback

For the use of the blocks to illustrate Sir Charles Parsons' lecture, the University Press, Cambridge, gratefully acknowledges the kindness of the Editors of *Engineering* and Mr Alex. Richardson, author of *The Evolution of the Parsons Steam Turbine*—a work which deals comprehensively with the subject.

Fig. 1. The "Turbinia" and the "Mauretania."

THE STEAM TURBINE

In modern times the progress of science has been phenomenally rapid. The old methods of research have given place to new. The almost infinite complexity of things has been recognized and methods, based on a co-ordination of data derived from accurate observation and tabulation of facts, have proved most successful in unravelling the secrets of Nature; and in this connection I cannot but allude to the work at the Cavendish Laboratory and also to that at the Engineering Laboratory in Cambridge, and to the association of Professor Ewing with the early establishment of records in steam consumption by the turbine.

In the practical sphere of engineering the same systematic research is now followed, and the old rule of thumb methods have been discarded. The discoveries and data made and tabulated by physicists, chemists, and metallurgists, are eagerly sought by the engineer, and as far as possible utilized by him in his designs. In many of the best equipped works, also, a large amount of experimental research, directly bearing on the business, is carried on by the staff.

The subject of our lecture today is the Steam Turbine, and it may be interesting to mention that the work was initially commenced because calculation showed that, from the known data, a successful steam turbine ought to be capable of construction. The practical development of this engine was thus commenced chiefly on the basis of the data of physicists, and, as giving some idea of the work involved in the investigation of the problem of marine propulsion by turbines, I may say that about £24,000 was spent before an order was received. Had the system been a failure or unsatisfactory, nearly the whole of this sum would have been lost.

Further, in order to prove the advantage of mechanical gearing of turbines in mercantile and war vessels about £20,000 has been recently expended, and considerable financial risks have been undertaken in relation to the first contracts.

With these preliminary remarks I now come to the subject of our lecture.

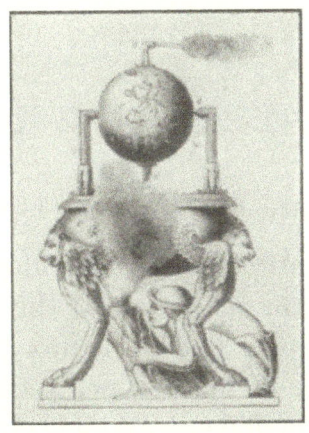

Fig. 2. Hero's Reaction Steam Wheel.

The first turbine of which there is any record was made by Hero of Alexandria, 2,000 years ago, and it is probably obvious

1—2

to most persons that some power can be obtained from a jet of steam either by the reaction of the jet itself, like a rocket or by its impact on some kind of paddle wheel. About the year 1837 several reaction steam wheels were made by Avery at Syracuse, New York, and by Wilson at Greenock, for driving circular saws and cotton gins. Fig. 3 shows the rotor of Avery's machine: steam is

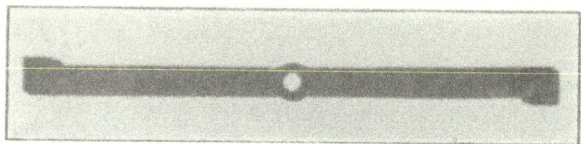

Fig. 3.   Rotor of Avery's Turbine.

introduced into it through a hollow shaft, and, by the reaction of the jets at the extremities, causes rotation. The rotor was 5 feet across, and the speed 880 feet per second. These wheels were inefficient, and it is not so obvious that an economical engine could be made on this principle. In the year 1888 Dr de Laval of Stockholm undertook the problem with a considerable measure of success. He caused the steam to issue from a trumpet-shaped jet, so that the energy of expansion might be utilized in giving velocity to the steam. Recent experiments have shown that in such

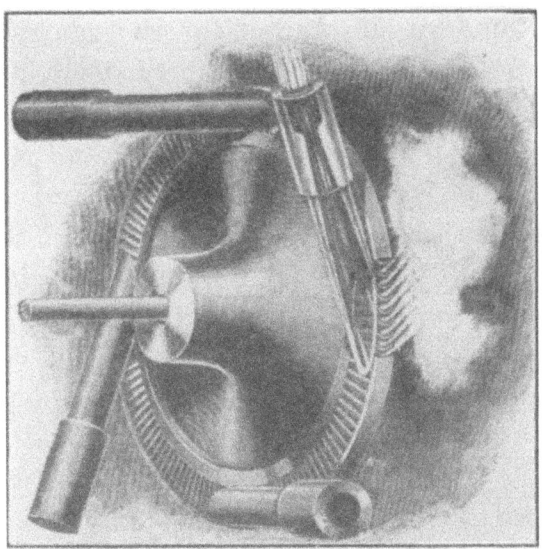

Fig 4   Dr de Lavals Turbine

jets about 80 per cent of the whole of the available energy in the steam is converted into kinetic energy of velocity in a straight line, the velocity attained into a vacuum being about 4,000 feet per second. Dr de Laval caused the steam to impinge on a paddle wheel made of the strongest steel, which revolved at the highest speed consistent with safety, or about half the velocity of a modern rifle bullet, for the centrifugal forces are enormous. Unfortunately, materials are not strong enough for the purpose, and the permissible speed of the wheel can only reach about two-thirds of that necessary for good economy, as I shall presently explain. Dr de Laval also introduced spiral helical gearing for reducing the enormous speed of rotation of his wheel (which needed to be kept of small diameter because of skin friction losses) to the ordinary speeds of things to be driven, and I shall allude to this gear later as a mechanism likely to play a very important part generally in future turbine developments.

In 1884 or four years previously, I dealt with the turbine problem in a different way. It seemed to me that moderate surface velocities and speeds of rotation were

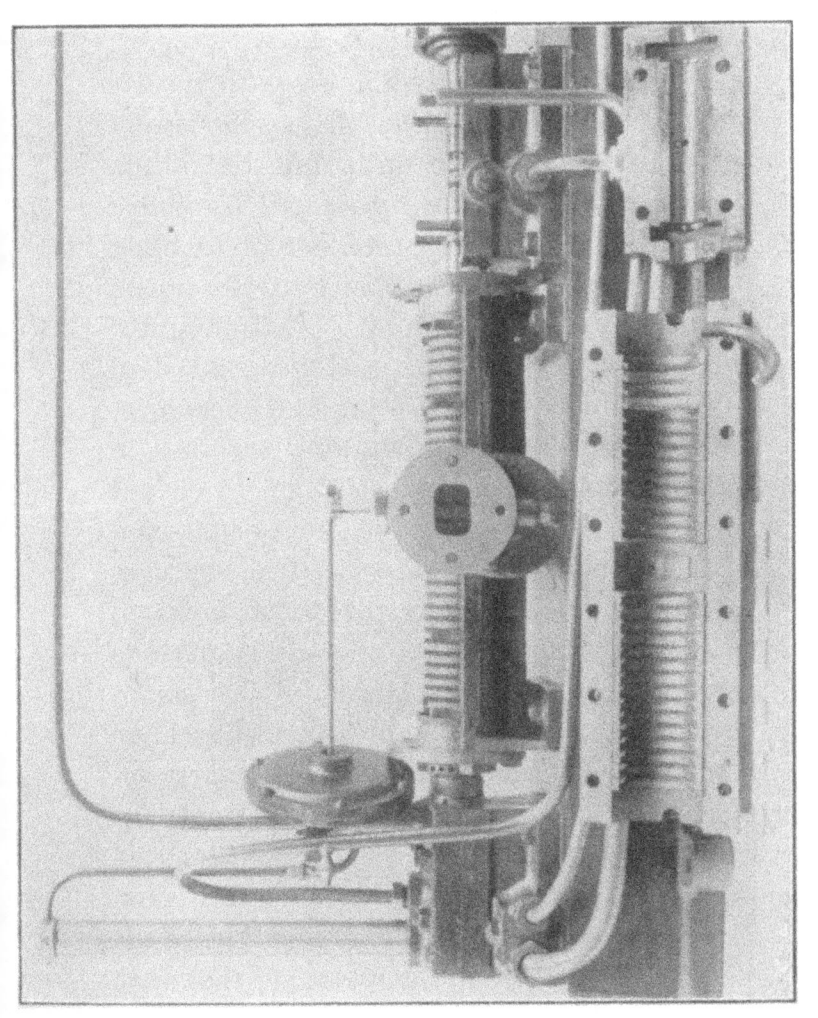

Fig 5 The First Parsons' Turbine

essential if the turbine motor was to receive general acceptance as a prime mover. I therefore decided to split up the fall in pressure of the steam into small fractional expansions over a large number of turbines in series, so that the velocity of the steam nowhere should be great. Consequently, as we shall see later, a moderate speed of turbine suffices for the highest economy. This principle of compounding turbines in series is now universally used in all except very small engines, where economy in steam is of secondary importance. The arrangement of small falls in pressure at each turbine also appeared to me to be surer to give a high efficiency, because the steam flowed practically in a non-expansive manner through each individual turbine, and consequently in an analogous way to water in hydraulic turbines whose high efficiency at that date had been proved by accurate tests.

I was also anxious to avoid the well-known cutting action on metal of steam at high velocity.

The close analogy between the laws for the flow of steam and water under small differences of pressure have been confirmed by experiment, and the usual formula of velocity $= \sqrt{2gh}$, where h is the hydraulic head, gives the velocity of issue from a jet for steam with small heads and also for water, and I shall presently follow this part of the subject further in dealing with the design of turbines. Having decided on the compound principle it was necessary to commence with small units at first; and thus, notwithstanding the compounding, the speed of revolutions though much reduced was still rather high.

The first compound steam turbine of 10 horse power (page 7) ran at 18,000 revolutions per minute, and had slightly elastic bearings to allow it to rotate about its dynamic or principal axis. The turbine teeth or blades were like a cog wheel, set at an angle and sharpened at the front edges, and the guide blades were similar. These are shown in Fig. 6 on the next page.

Gradually the form of the blades was

Figs. 7 to 11.   Blades and Distance Pieces tried in 1894

DUMMY STRIPS

Figs. 16 and 17.   Successive Types of Dummy Strips.

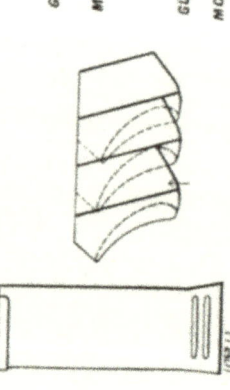

GUIDE BLADES.

MOVING BLADES.

GUIDE BLADES

MOVING BLADES

Figs. 12 to 15.   Blades experimented with in 1900.

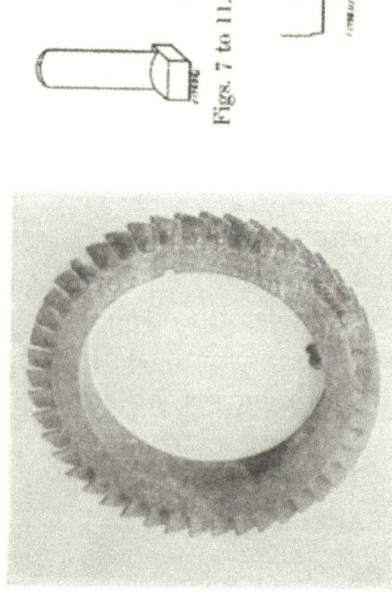

Fig. 6.   Straight Blading of First
Turbine made, 1884.

Figs. 6 to 17.   Systems of Blading.

improved as a result of experiments and some of these are shown on page 10. Curved blades with thickened backs were introduced. The blades were cut off to length from brass, hard rolled and drawn to the required section, and inserted into a groove with distance pieces between and caulked up tightly.

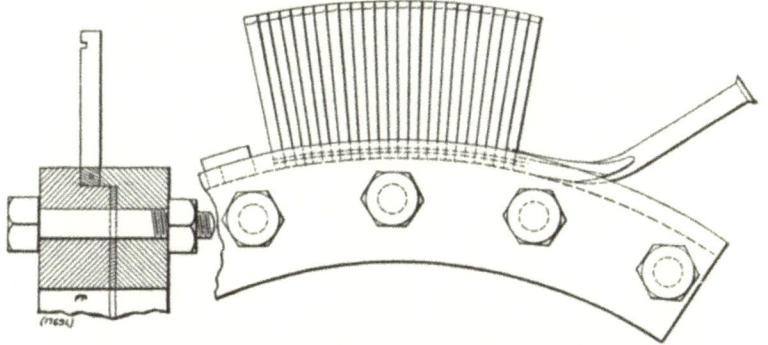

Figs. 18 and 19. Formers for making Segments of Blades.

Dummy labyrinth packings of various types were introduced. Two of these are illustrated in Figs. 16 and 17 on page 10. The design was improved, generally, so as to reduce steam leakages and to provide for greater ratios of expansion.

Fig 20   Forming Segments of Casings and Rotor Blading

The diagrams on page 11 show the latest method of forming segments of blades by stringing blades and distance pieces alternately on wire within a groove formed of two castings bolted together and corresponding to the groove of the turbine rotor or casing. The engraving on page 12 illustrates these segments being made. The view on page 15 shows segments being fitted in a rotor.

I have said that steam behaves almost like an incompressible fluid in each turbine of the series, but because of its elasticity its volume gradually increases with the succession of small falls of pressure, and the succeeding turbines consequently are made larger and larger. This enlargement is secured in three ways: (1) by increasing the height of blade, (2) by increasing the diameter of the succeeding drums, and (3) by altering the angles and openings between the blades. All three methods are generally adopted (page 17) to accommodate the expanding volume of the steam which in a condensing turbine reaches one hundredfold or more before it issues from the last blades to the condenser.

Now as to the best speed of the blades, it will be easily seen that in order to obtain as much power as possible from a given quantity of steam, each row must work under appropriate conditions. This has been found by experiment to require that the velocity of the blades relatively to the guide blades shall be from one-half to three-quarters of the velocity of the steam passing through them, or more accurately equal to one-half to three-quarters of the velocity of issue from rest due to the drop of pressure in guides or moving blades, for in the usual reaction turbine the guides are identical with the moving blades.

The curve for efficiency in relation to the velocity ratio has a fairly flat top, so that the speed of the turbine may be varied considerably about that for maximum efficiency without materially affecting the result.

In compound land turbines the efficiency of the initial rows is about 60 per cent., and

Fig 21   Blading a Rotor at the Turbinia Works

of the latter rows 75 to 85 per cent., and considering the whole turbine, approximately 75 per cent. of the energy in the steam is delivered on to the shaft. The expansion curve of the steam lies between the adiabatic and isothermal curves, but nearer the former, because 75 per cent. is converted into work on the shaft and only 25 per cent. is lost by friction and eddies in the steam and therefore converted into heat.

In turbine design the expression of the velocity ratio between the steam and blades may be represented by the integral of the square of the velocity of each row through the turbine, and if, for instance, this integral is numerically equal to 150,000,—a usual allowance for land turbines,—then we know that, with a boiler pressure of 200 lbs. and a good vacuum, the velocity of the blades will be a little over one half that of the steam, and the turbine will be working close up to that speed which gives the maximum efficiency. In large marine turbines where weight and space are of importance the integral may be

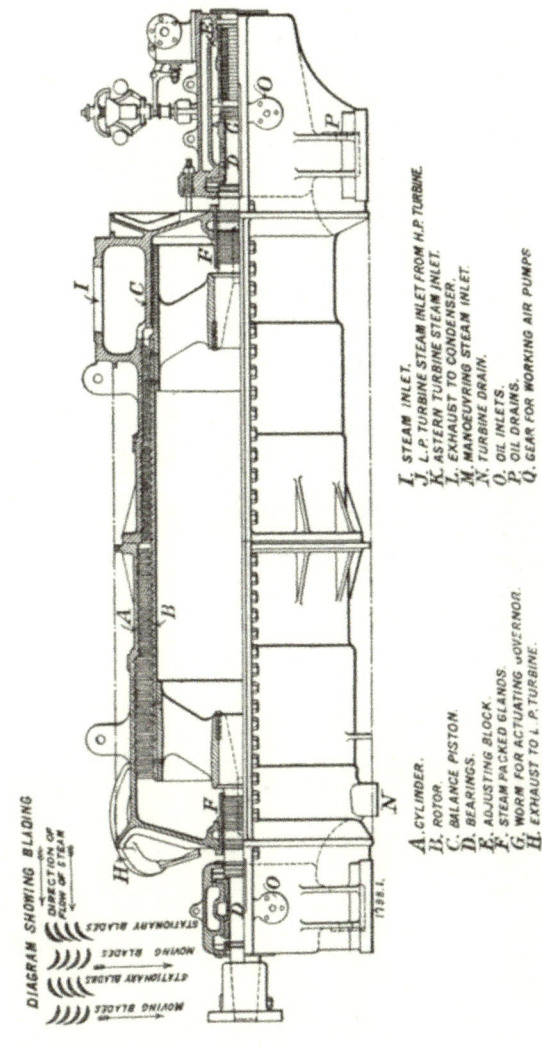

DIAGRAM SHOWING BLADING

DIRECTION OF FLOW OF STEAM

MOVING BLADES
STATIONARY BLADES
MOVING BLADES
STATIONARY BLADES

A. CYLINDER.
B. ROTOR.
C. BALANCE PISTON.
D. BEARINGS.
E. ADJUSTING BLOCK.
F. STEAM PACKED GLANDS.
G. WORM FOR ACTUATING GOVERNOR.
H. EXHAUST TO L.P. TURBINE.

I. STEAM INLET.
J. L.P. TURBINE STEAM INLET FROM H.P. TURBINE.
K. ASTERN TURBINE STEAM INLET.
L. EXHAUST TO CONDENSER.
M. MANOEUVRING STEAM INLET.
N. TURBINE DRAIN.
O. OIL INLETS.
P. OIL DRAINS.
Q. GEAR FOR WORKING AIR PUMPS

Fig. 22. Section of High-Pressure Ahead Marine Turbine.

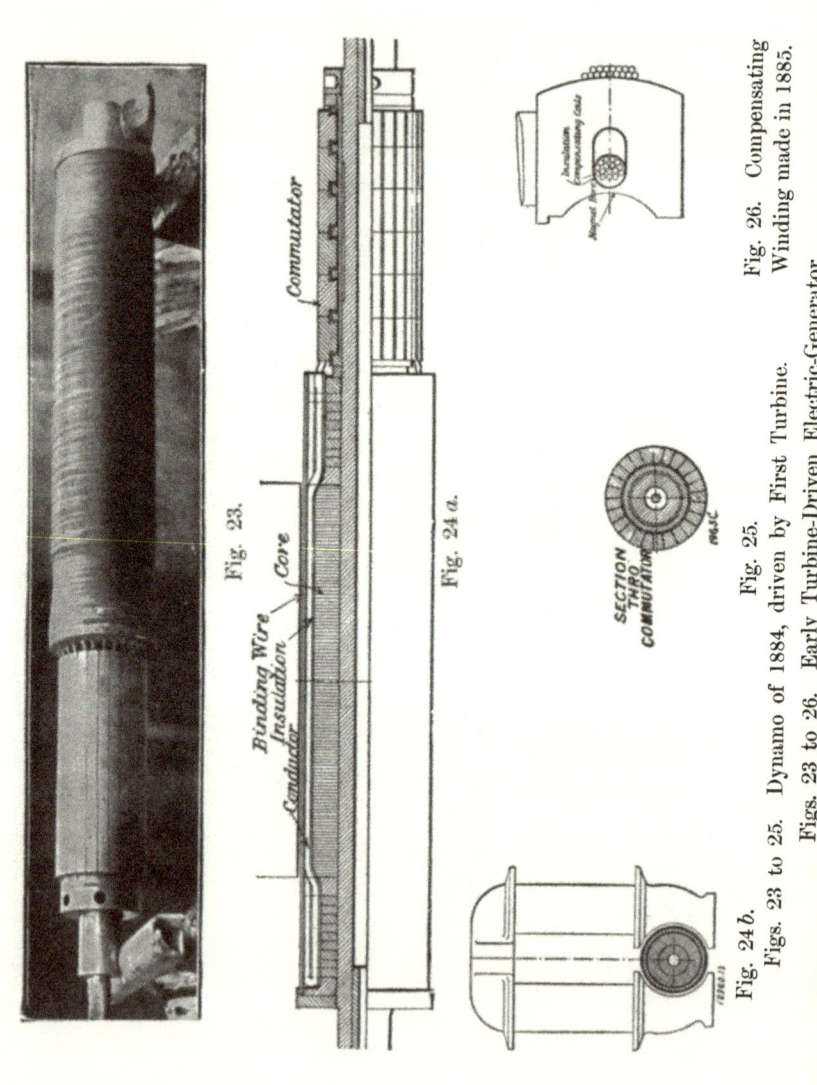

Fig. 23.

Binding Wire
Insulation
Conductor
Core
Commutator

Fig. 24 a.

Fig. 24 b.

SECTION THRO COMMUTATOR

Fig. 25.

Insulation
Compensating Cable

Fig. 26. Compensating
Winding made in 1885.

Figs. 23 to 25. Dynamo of 1884, driven by First Turbine.

Figs. 23 to 26. Early Turbine-Driven Electric-Generator.

from 80,000 to 120,000 or more. With the first figure a loss of efficiency of about 10 per cent. below the highest attainable is accepted, and with the latter figure the deficit is only about 3 per cent.

The construction of a suitable dynamo to run with the turbine involved nearly as much trouble as the turbine itself: the chief features were the adoption of very low magnetic densities in the armature core and small diameters and means to resist the great centrifugal forces as shown in the views on page 18. The dynamo was also mounted in elastic bearings. Now that the turbine has found its most suitable field in large powers to which we always looked forward and as the speed of revolution has been consequently reduced, elasticity in the bearings is less essential, and in large land plants and in marine work rigid bearings are universal.

There are many forms of turbines on the market. It is only necessary, however, for us here to consider the four chief types which are:—

First, the compound reaction turbine with which we have been dealing, representing over 90 per cent. of all marine turbines in use in the world, and about half the land turbines driving dynamos.

Second, the de Laval, which is only used for small powers.

Third, the "multiple impulse, compounded" or Curtis, which has been chiefly used on land, but which has been fitted in a few ships.

Lastly, a combination of the compound reaction type with one or more "multiple impulse or Curtis elements" at the high-pressure end to replace the reaction blading.

We may dismiss the other varieties as simply modifications of the original types without possessing any originality or scientific interest.

Now let me further explain the multiple impulse type, and commence by saying that it is the only substantial innovation in turbine practice since the compound reaction and the de Laval turbines came into use.

It was proposed by Pilbrow in 1842, and first brought into successful operation by Curtis in 1896. Some consideration should be given to it as involving several characteristic points of difference from what has been said about the compound reaction type. Curtis in the first place used the de Laval divergent

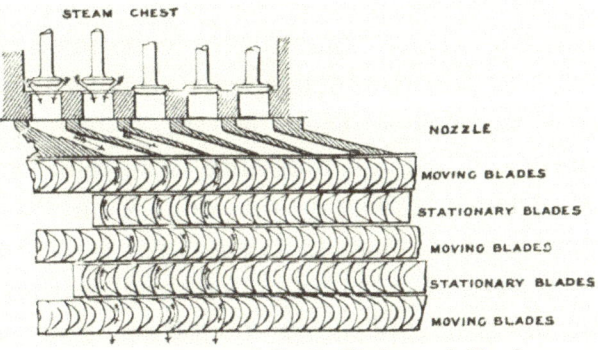

Fig. 27. Diagram of Curtis Blades and Nozzles.

nozzle, and he also used compounding to the limited extent of only 5 to 9 stages, as compared with 50 to 100 in the compound type. With these provisos the same principles in the abstract as regards velocity ratio now apply, and the steam issuing from the jets rebounds again and again between the fixed and moving buckets at each velocity compounded stage: the best velocity ratio in a four row multiple impulse is only one-seventh and the efficiency about 44 per cent., and therefore much lower than that of reaction blading, which as we have stated is under favourable conditions 75 to 85 per cent.

The advantages, however, to be derived from the use of some multiple impulse elements at the commencement of the turbine are that because there is very little loss in them from leakage, therefore in spite of their low intrinsic efficiency, one or more multiple impulse wheels can in certain cases usefully replace reaction blading. The explanation is that in turbines of the compound reaction type of moderate power and slow speed of revolution the blades are often very short at the commencement, and consequently there is in such cases excessive loss by leakage through the clearance space, which brings the efficiency below that of impulse blading. In most cases one multiple impulse wheel is preferred, followed by reaction blading. Such

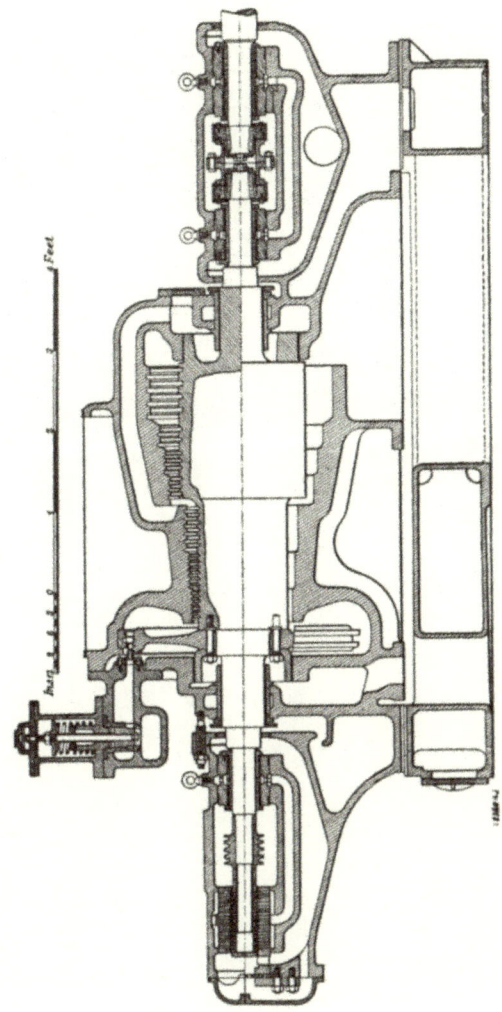

Fig. 28. Parsons' Combined Impulse-Reaction Turbine.

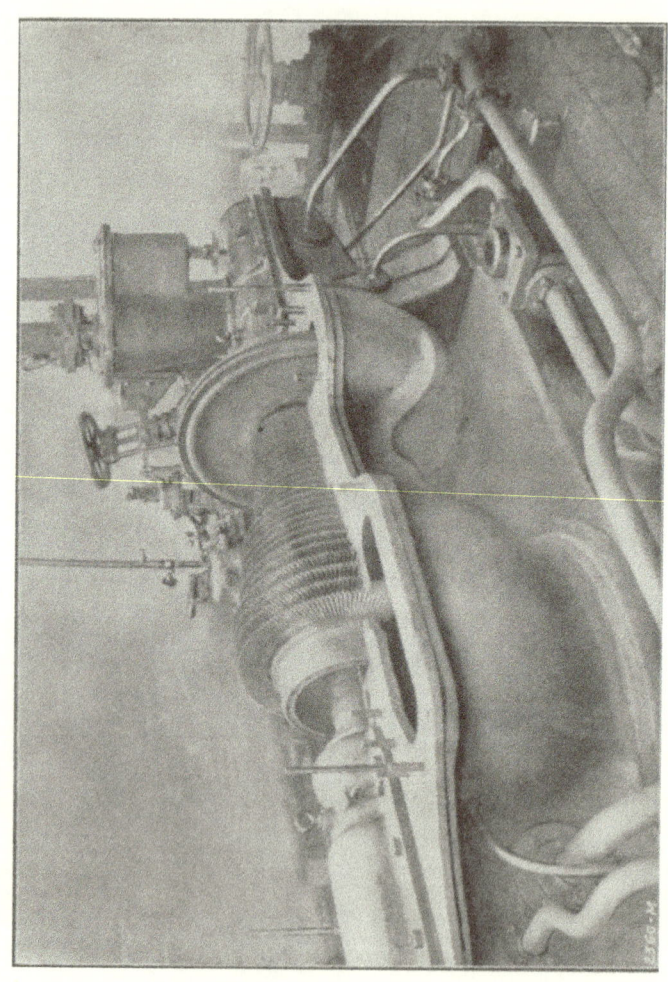

Fig 29 Parsons' Combined Impulse-Reaction Turbine

impulse-reaction turbines are illustrated on pages 23 and 24.

When highly superheated steam is used the temperature is much reduced by expansion in the jets and work done in the impulse wheel before it passes to the main turbine casing.

The highest efficiency yet attained by land turbines has, however, been with the pure compound reaction type of large size, where the high pressure portion is contained in a separate casing of short length and great rigidity, now made usually of steel. The working clearances can by this arrangement be reduced to a minimum and the highest efficiency attained.

The first turbine imported into Germany in 1900, of 2000 H.P., was on this principle, while the latest turbines are of 12,000 H.P., and generate current for the Metropolitan Railway in London.

In marine work the same principle has been almost universal since 1896, when the original single turbine of the "Turbinia" was

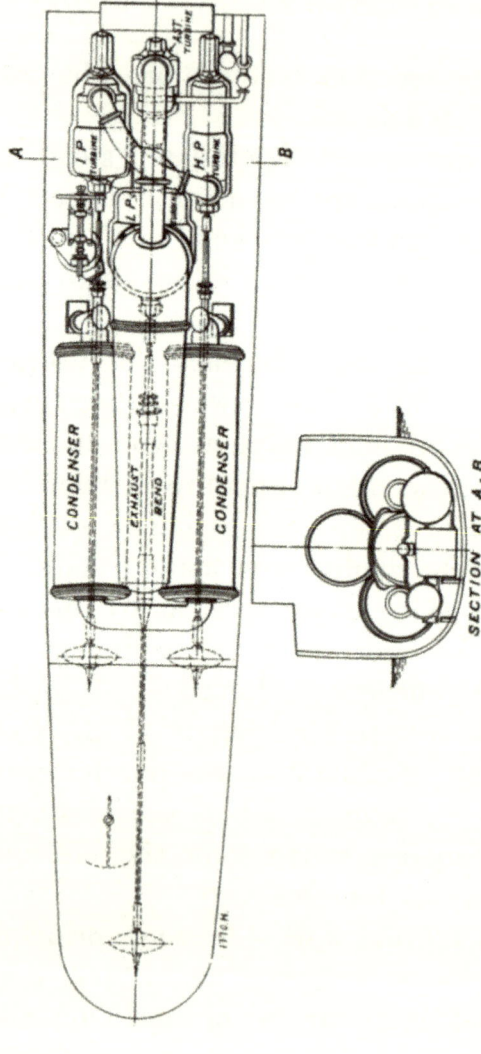

Figs. 30 and 31. General arrangement of Turbines in Series in the "Turbinia."

CONDENSER

EXHAUST
BEND

CONDENSER

I.P.
TURBINE

AFT
TURBINE

H.P.
TURBINE

L.P.

A

B

SECTION AT A.B
LOOKING AFT

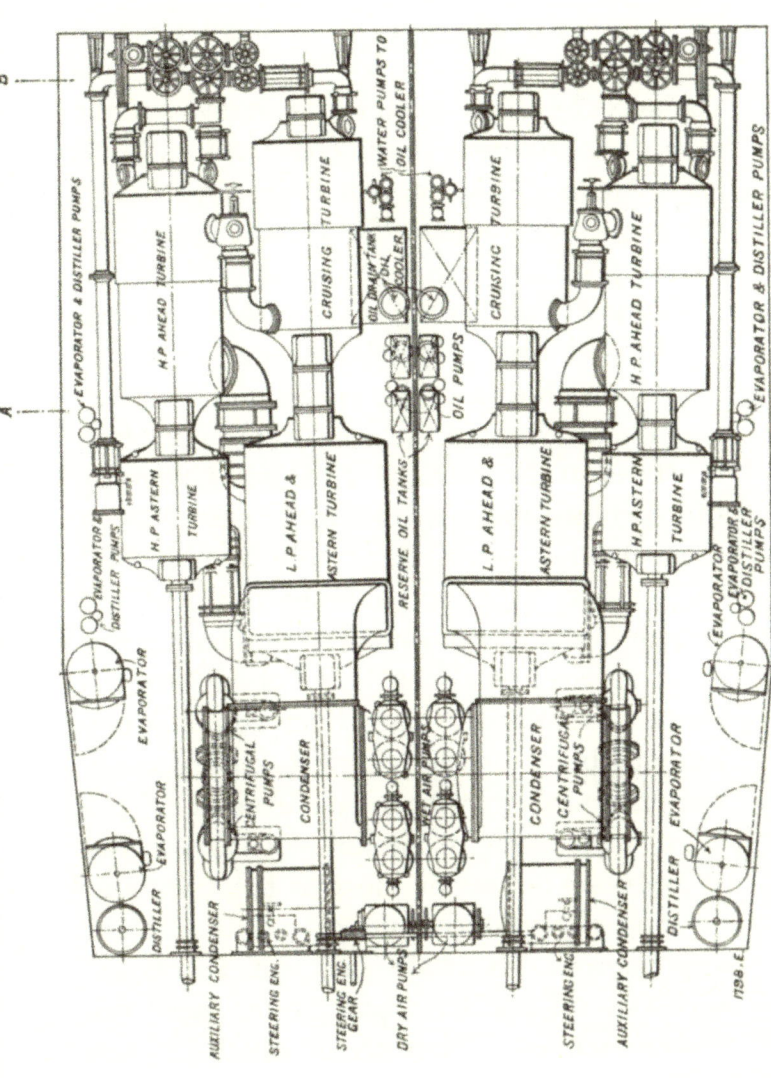

Fig. 32. Turbines in Series for Four Shaft arrangement in Battleships.

replaced by three turbines in series (on the steam) on different shafts (page 26), and it is adopted in all the largest liners and almost all large war vessels. In marine work this division of the turbine has the additional advantage that owing to the power being subdivided over three shafts, smaller screws are admissible, and the speed of revolution may be increased in the case of three turbines in series in the ratio of 1 to $\sqrt{3}$. Generally the turbines are placed two in series, as in cross-channel boats, the " Mauretania " and " Lusitania," torpedo craft, battleships, and cruisers (page 27), or sometimes three in series (page 29) as in the liner " La France "and the latest and largest Cunard liner now building. Four turbines in series have been proposed, but have not as yet been constructed.

A war vessel in commission is working at reduced power for most of the time, and on long voyages economy of fuel is of great importance. To attain this end, additional turbines are fitted in front of the main full power turbines. They are of small size, and

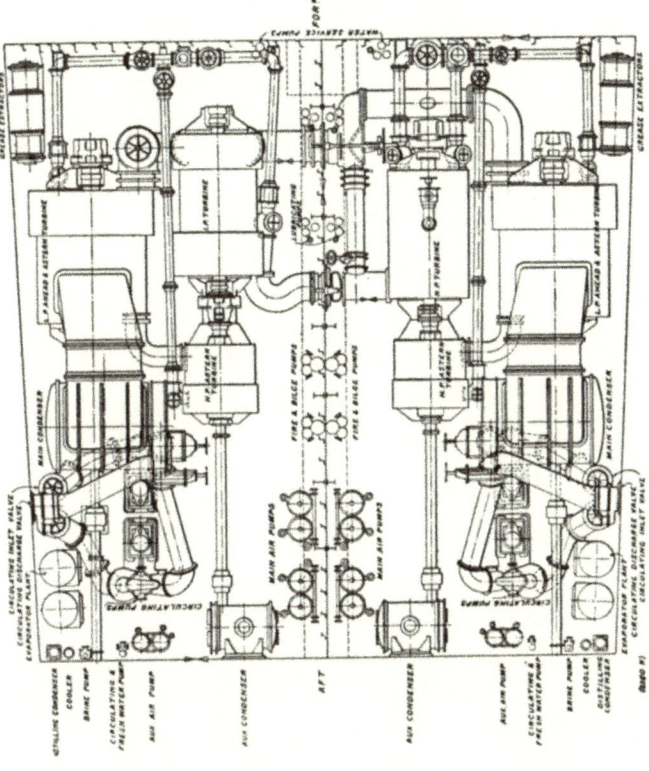

Fig 33   Turbines arranged, three in series, for a Four Screw Battleship

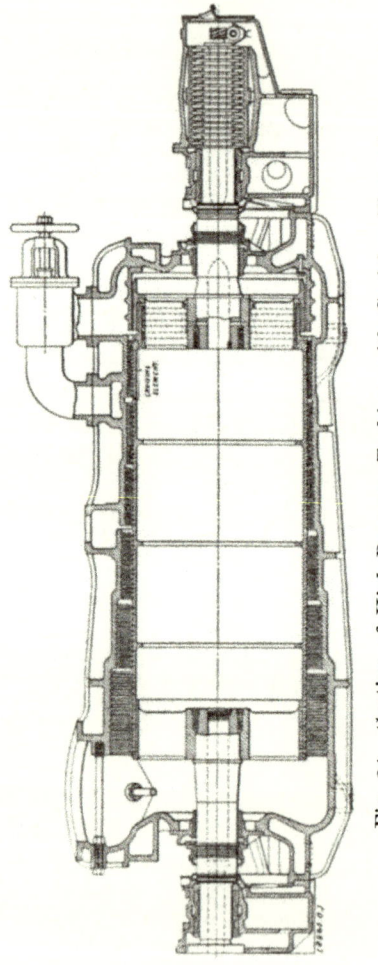

Fig. 34. Section of High-Pressure Turbine, with Cruising Element.

in separate casings, or they may form an integral portion of the main high pressure turbine, which is then lengthened by the addition of the cruising portion (page 30). They are partially by-passed as more power is required, and at full speed they are entirely by-passed, or, when in separate casings, are completely isolated from the steam supply by suitable valves, and are generally connected to the condenser and rotate in vacuum, so that there is no appreciable resistance to rotation. In some instances of modern naval construction one or more multiple impulse wheels have constituted the cruising element.

Before passing to the consideration of other applications of the turbine I should like, with your permission, to repeat an experiment which illustrates the phenomenon of cavitation. The chief difficulty in applying the turbine to marine propulsion arose in the breaking away of the water, or the hollowing out of vacuous cavities when it was attempted to rotate the screw above certain limits. The phenomenon was first observed by Sir John

Thornycroft and Mr Sydney Barnaby. They designated this phenomenon by the appropriate name "Cavitation," and it entails, by

Fig. 35. Apparatus for Experiments on Cavitation.

the way, a great loss of power. The remedy lies in using very wide blades covering about

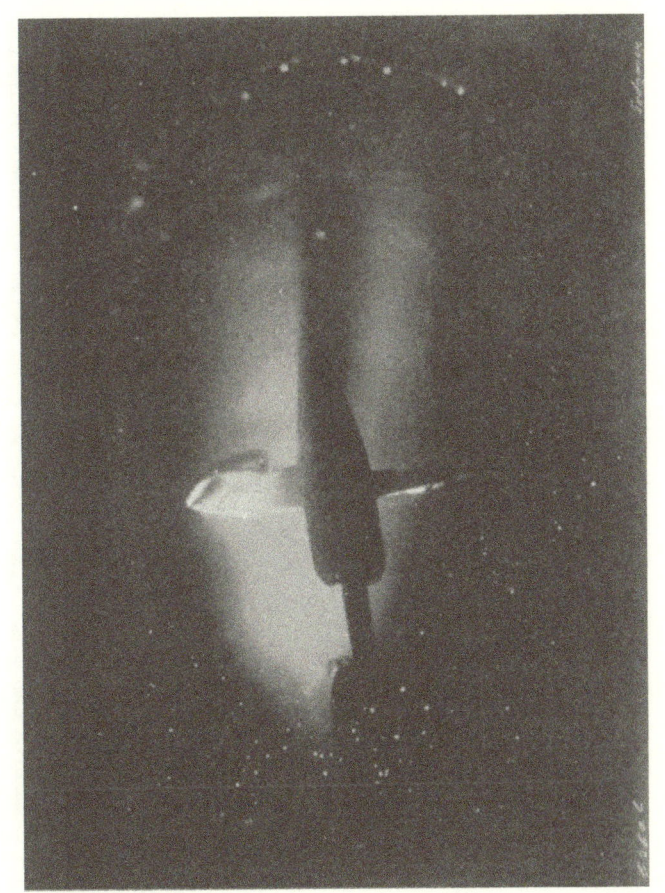

Fig. 36. Beginning of Cavitation ; 1500 Revolutions.

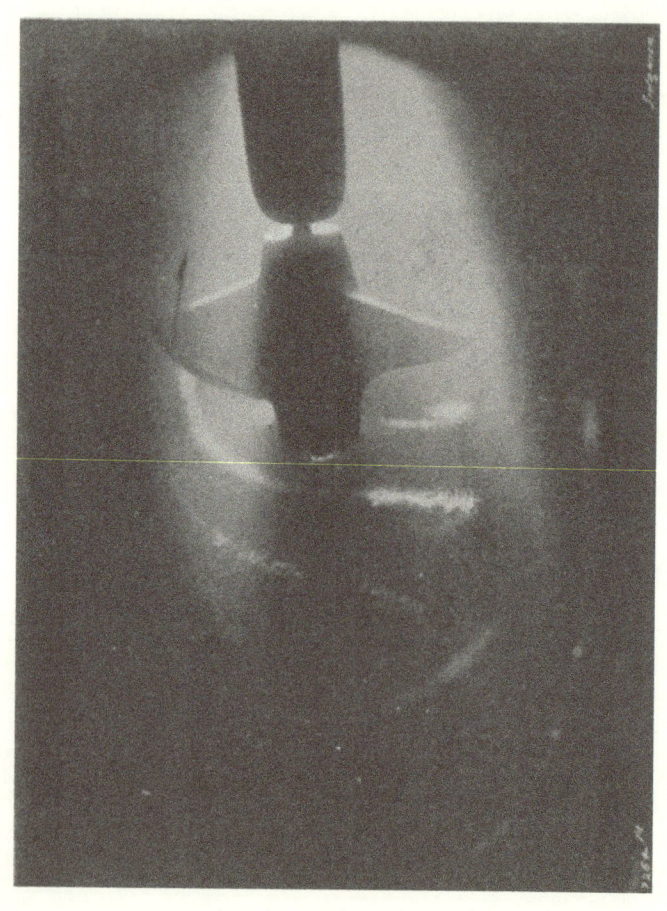

Fig. 37. Advanced stage in Cavitation ; 1800 Revolutions.

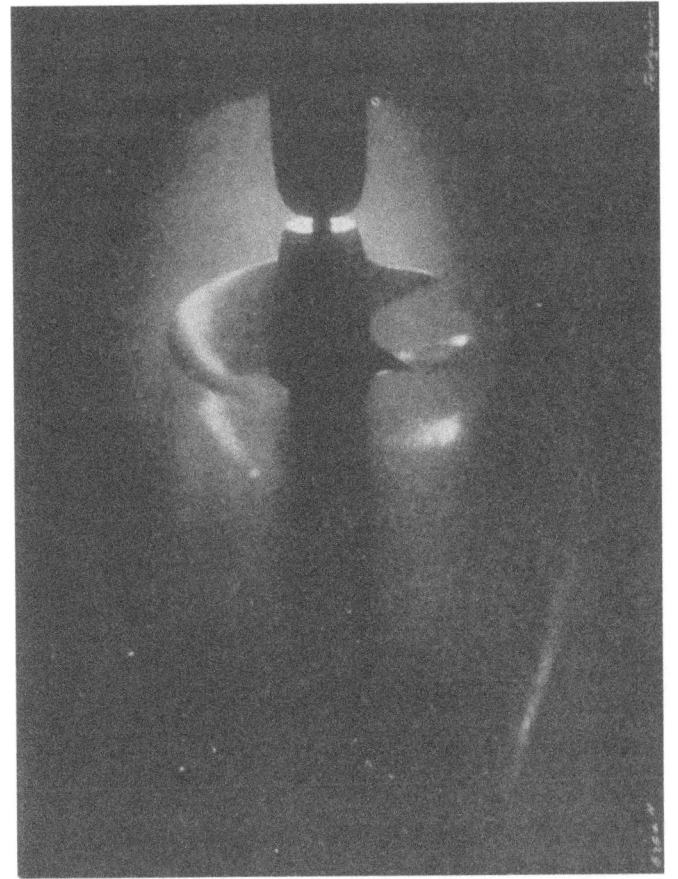

Fig. 38. Final stage of Cavitation ; 2000 Revolutions.

two-thirds of the disc area of the propeller, so as to present a very large bearing surface on the water, and this expedient effectually prevents its giving way under the force necessary to propel the vessel.

In models, and in vessels of moderate speed, the forces are not sufficient to tear

TABLE I. *Performance of Parsons' Turbo-Generators at Different Epochs.*

Date	Power	Steam per kw. hour	Vacuum (Bar. 30″)	Superheat	Steam pressure per sq. inch
	kw.	lb.	in.	Deg. F.	lb.
1885	4	200	0*	0*	60
1888	75	55	0*	0*	100
1892	100	27·00	27	50	100
1900	1250	18·22	28·4	125	130
1902	3000	14·74	27	235	138
1907–1910	5000	13·2	28·8	120	200

* These were non-condensing turbines using saturated steam.

the water asunder, but if the pressure of the atmosphere is removed by an air pump, a model screw will cavitate at a comparatively moderate speed.

The improvement in efficiency resulting from the successive modifications and improvements in the proportions of turbines, and also arising from the increase in the size is shown by the particulars given in the Table opposite.

Table II on this page gives corresponding data in regard to marine turbines.

TABLE II. *Performance of Notable Ships of Different Epochs with Parsons' Turbines.*

Date	Name of Ship	Length	Displacement	H P	Steam consumption per SHP. per hr. for all purposes	Speed in knots
		ft.	tons		lb	
1897	" Turbinia "	100	44½	2,300	15	32·75
1901	"King Edward"	250	650	3,500	16	20·48
1905	H.M S. " Amethyst "	360	3,000	14,000	13·6	23·63
1906	H.M.S. "Dreadnought"	490	17,900	24,712	15·3	21·25
1907	"Mauretania" & "Lusitania"	785	40,000	74,000	14·4	26·0

Many warships are now being fitted with installations with double and treble turbines in series on the steam and exceeding the

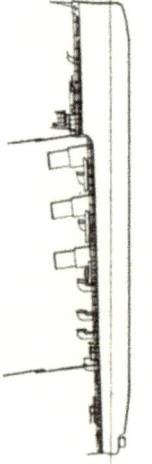

"Turbinia," built 1894 ; length, 100 ft., displacement, 44½ tons ; horse-power, 2300 , speed, 32¾ knots.

H. M. Torpedo-Boat Destroyer "Velox," built 1902, length, 210 ft. ; displacement, 420 tons ; shaft horse-power, 8000 , speed, 27.12 knots.

H. M. Torpedo-Boat Destroyer "Eden," built 1903 ; length, 220 ft. ; displacement, 540 tons ; shaft horse-power, 7000 ; speed, 26 22 knots.

H. M. Torpedo-Boat Destroyer "Swift," built 1908 ; length, 345 ft. ; displacement, 2170 tons ; shaft horse-power, 30,000, speed. 35.3 knots.

H. M. 3rd Class Cruiser " Amethyst," built 1905 ; length, 360 ft. ; displacement, 3000 tons ; shaft horse-power (estimated), 14,200 ; speed, 23.63 knots.

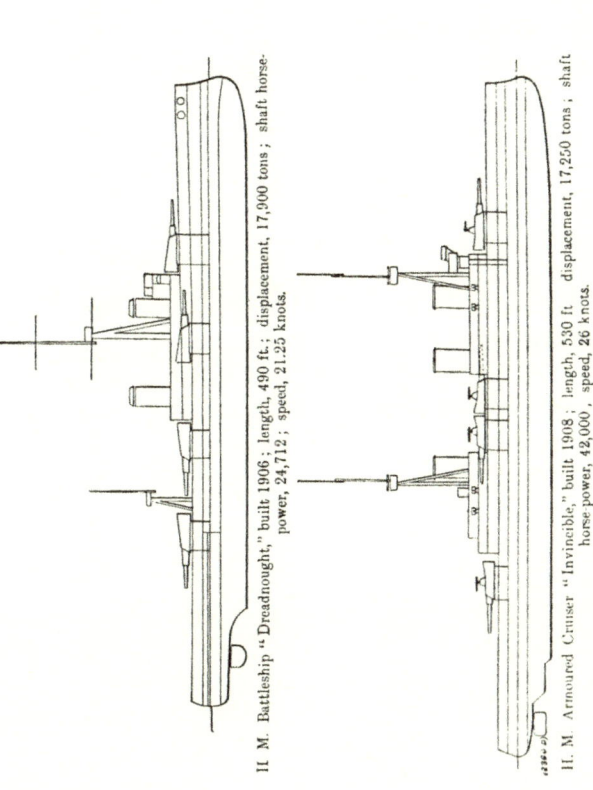

H. M. Battleship "Dreadnought," built 1906; length, 490 ft.; displacement, 17,900 tons; shaft horse-power, 24,712; speed, 21.25 knots.

H. M. Armoured Cruiser "Invincible," built 1908; length, 530 ft.; displacement, 17,250 tons; shaft horse-power, 42,000; speed, 26 knots.

Fig. 39. Diagram showing growth in size between 1894 and 1908 of Turbine-propelled Warships.

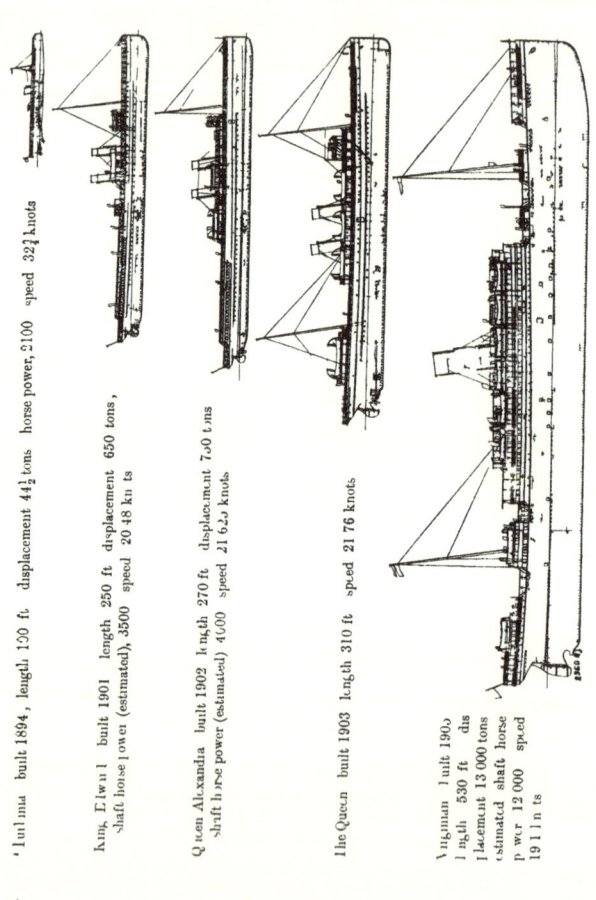

Turbinia built 1894, length 100 ft displacement 44½ tons horse power, 2100 speed 32¼ knots

King Edward built 1901 length 250 ft displacement 650 tons, shaft horse power (estimated), 3500 speed 20·48 knots

Queen Alexandra built 1902 length 270 ft displacement 760 tons shaft horse power (estimated) 4000 speed 21·625 knots

The Queen built 1903 length 310 ft speed 21·76 knots

Virginian built 1905 length 530 ft displacement 13 000 tons estimated shaft horse power 12 000 speed 19·11 knots

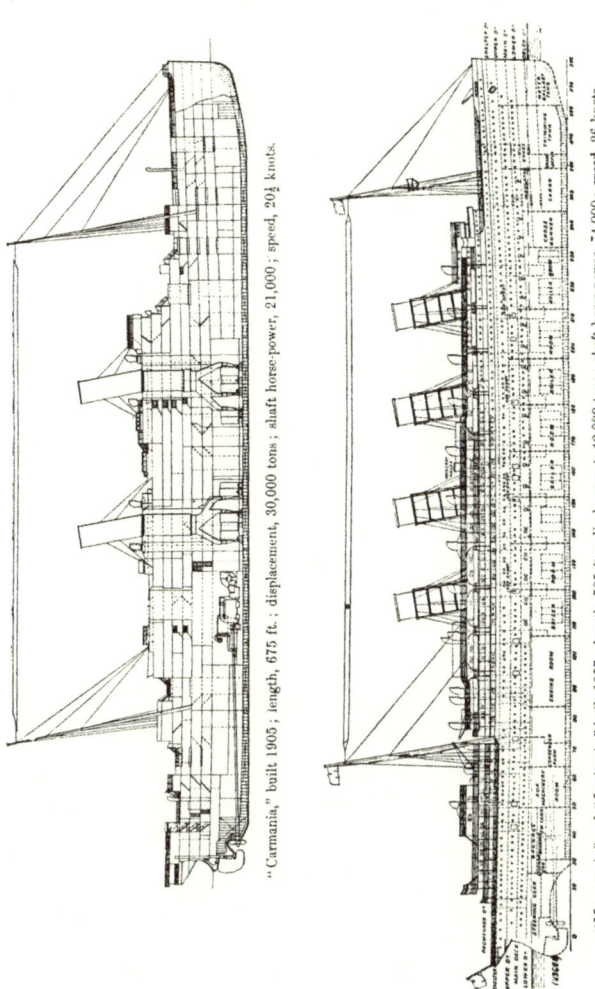

"Carmania," built 1905 ; length, 675 ft. ; displacement, 30,000 tons ; shaft horse-power, 21,000 ; speed, 20½ knots.

"Mauretania " and "Lusitania," built 1907 ; length, 785 ft. ; displacement, 40,000 tons ; shaft horse-power, 74,000 ; speed, 26 knots.

Fig. 40. Diagram showing growth in size of Turbine-propelled Merchant Ships.

power developed by the "Mauretania" and "Lusitania." The aggregate power of Parsons' turbines fitted for marine propulsion up to date is six million shaft horse-power. This embraces ships of practically all nationalities. The power of turbines of the same type made for generating electricity and other duty on land is also about six million shaft horse-power—considerably more than the available power of Niagara Falls. The diagrams on the four preceding pages indicate the size of ships fitted at different successive periods for the Royal Navy and for the merchant marine.

The marine turbine, with the modifications we have so far described, is only suitable for vessels of over 16 knots speed, and to extend its use to vessels of a less speed than this, which comprise two-thirds of the tonnage of the world, has been our constant aim. The first plan for the attainment of this end is somewhat in the nature of a compromise, and is called the combination system, because the reciprocating engine is used to take the first

part of the expansion and the turbine the
last. From what we have said it will be
apparent that this coalition of the reciprocating engine and turbine is a good one,
because each works under favourable conditions. The reciprocating engine expands

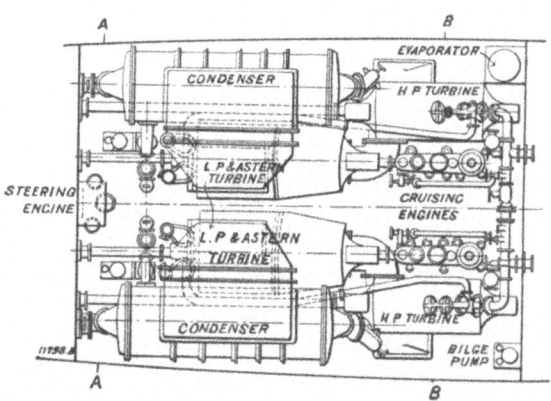

Fig. 41. First set of Combination Machinery—in H.M.S. "Velox."

the steam to about atmospheric pressure,
and the turbine carries on the expansion
with high efficiency down to the pressure in
the condenser. Now, though a large and
high speed turbine can be made to deal with
the high pressure portion of the expansion

Fig 42   The First Turbine Commercial Steamer—The " King Edward."

as economically as a reciprocating engine, a slow speed turbine cannot be made to do so, but on the other hand a slow speed turbine expands low pressure steam much further and more economically than any reciprocating engine. Under this system the turbine generally is made to develop about one-third of the whole power.

About 15 years ago this plan was worked out and the British Admiralty destroyer " Velox " was so fitted in 1902 (page 43), but no further practical steps were taken towards its application until about three years ago.

Messrs Denny of Dumbarton, who in 1901 built the first mercantile turbine vessel, the " King Edward," in 1908 built the first combination vessel, the "Otaki" of 9,900 tons and 13 knots speed. She has ordinary twin screws driven by triple expansion engines which exhaust into a turbine driving a central screw as illustrated on the next page. The initial pressure at the turbine is 9 lbs. absolute, and it develops one-third of the whole power. This combination vessel was

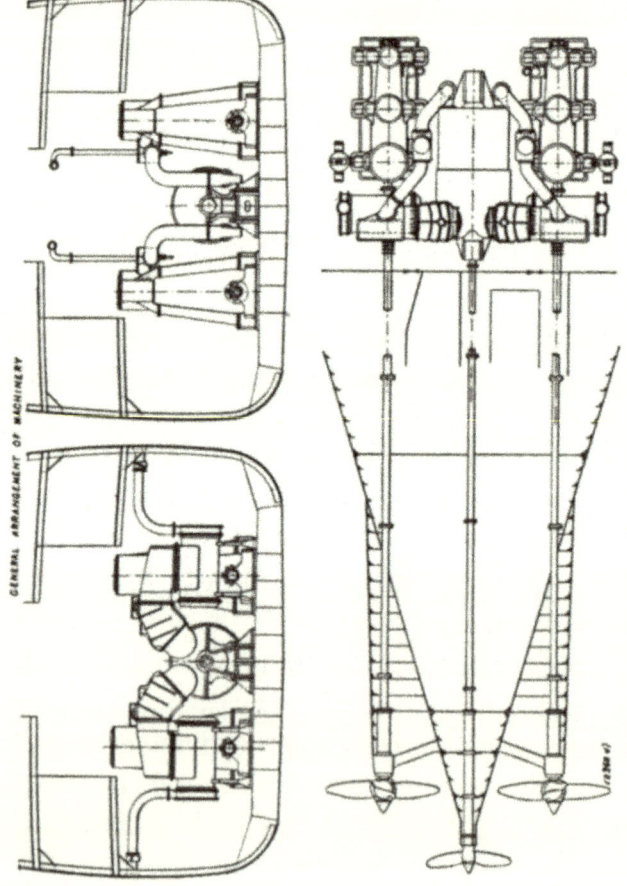

GENERAL ARRANGEMENT OF MACHINERY

Figs. 43 to 45.   The Combination Machinery in the "Otaki."

found to consume 12 per cent. less coal than her sister vessel on the same service,—the "Orari," fitted with quadruple reciprocating engines. The next combination vessel was the "Laurentic" of 20,000 tons built by Messrs Harland and Wolff, a sister vessel to the "Megantic," fitted with quadruple engines, and on service at the same speed the saving in coal by the combination is 14 per cent.

The combination system has also been adopted in the White Star liners "Olympic" and "Titanic" of 60,000 tons displacement, as well as in some other vessels at home and abroad.

There is another promising solution of the problem for applying the turbine to slower vessels, which will extend its field still further over that of the reciprocating engine. I mentioned before that de Laval had in the 'eighties introduced helical tooth gear for reducing the speed of his little turbines. For 23 years it has worked admirably on a small scale. Recent experiments, however, have led to the assurance of equal success on a large scale for the transmission of large powers.

Preliminary experiments were made some years ago on helical reduction gear, which showed a mechanical efficiency of over 98 %, and a 22 feet launch was constructed in 1897 : the working speed of the turbine was 20,000 revolutions per minute, which was geared in one reduction of 14 to 1 to the twin screws. The speed attained was 9 miles per hour, and this little boat was many years in use as a yacht's gig. She was the first geared turbine vessel. The next step was to test geared turbines in a typical cargo boat, and the "Vespasian" was purchased in 1908. She is of 4,350 tons displacement, and was propelled by a good triple expansion engine of 900 horse-power. After thoroughly overhauling and testing her existing machinery for coal and water consumption, to make sure that it was in thorough good order, the engine was taken out and replaced by geared turbines, the propeller, shafting, and boilers remaining the same. On again testing for

economy with the new machinery a gain
of 15 per cent. was shown over the reciprocating engine, and a subsequent alteration

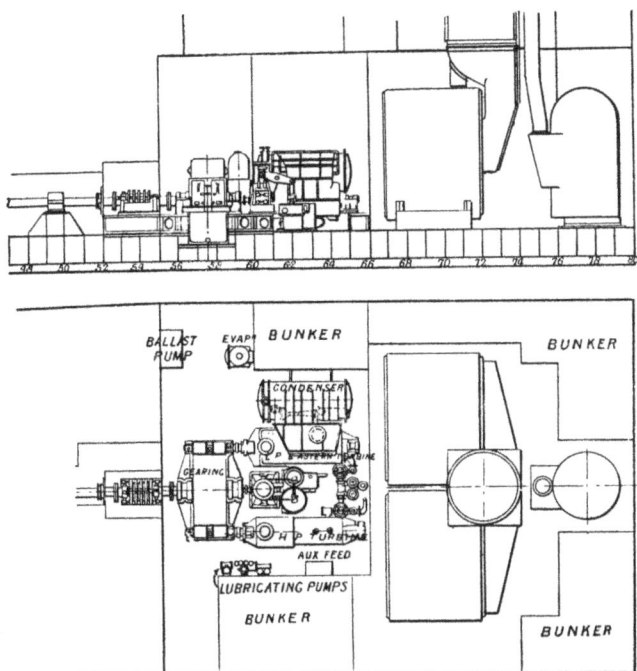

Figs. 46 and 47. The Geared Turbines in the "Vespasian."

to the propeller has increased this gain to
22 per cent., a very remarkable saving. The
new machinery, which is much lighter than

the old, consists of a high pressure and a low pressure turbine, each driving a pinion at 1,400 revolutions, gearing into a spur wheel on the screw shaft making 70 revolutions per minute (page 52). The gearing is entirely enclosed in a casing, and is continually sprayed with oil by a pump.

It is interesting to compare the working of the new and the old machinery. The appended diagram (Fig. 48) shows the comparative water consumption with reciprocating and turbine engines. Everyone who has experienced a rough sea in a screw vessel knows the disagreeable sensation of the racing of the engines whenever the screw comes out of the water. In the turbine vessel nothing of the kind occurs, and the reason is very simple. It is because of the great angular momentum of turbines, which is about 50 times that of ordinary engines, consequently they gather speed so slowly that before they have appreciably accelerated the screw is down again in the water. Ordinary engines often accelerate up to

three times their ordinary speed in a heavy
sea, and what shakes the ship and breaks

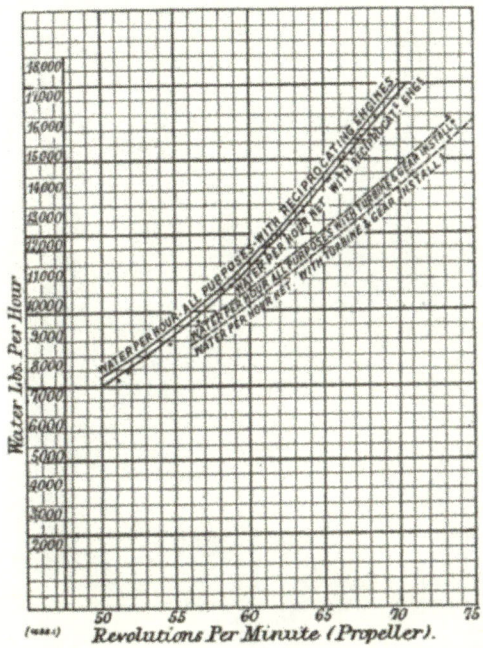

Fig. 48. Water Consumption of "Vespasian" in service with
Reciprocating Engines and with Geared Turbines, the propeller being the same in both cases.

the screw shafts is the shock on the plunging
of the madly whirling propeller into the sea.

The "Vespasian" has now covered 20,000

4—2

Fig 49 View of Gearing in "Vespasian"

miles in all weathers and carried 90,000 tons
of coal from Newcastle to Rotterdam.

The pinion on the table was removed
from the vessel a month ago for the purpose
of showing it at this lecture. As you can see
it shows no sign of wear. The gearing is
illustrated on page 52.

Gearing promises to play a very important part in war vessels for increasing
the economy at reduced speeds. I ex-
plained the difficulty in obtaining good
economy under such conditions, and by means
of geared high speed turbines the efficiency
will be greatly increased. The Turbinia
Company are now constructing two 30-knot
destroyers of 15,000 horse-power with this
arrangement (page 54). The high pressure
portion and cruising elements are geared in
the ratios of 3 to 1 and 5 to 1 respectively
to the main low pressure direct coupled
turbine, and their use will increase the effec-
tive radius of action at cruising speed by
nearly 50 per cent. over that of a similar
destroyer without gearing.

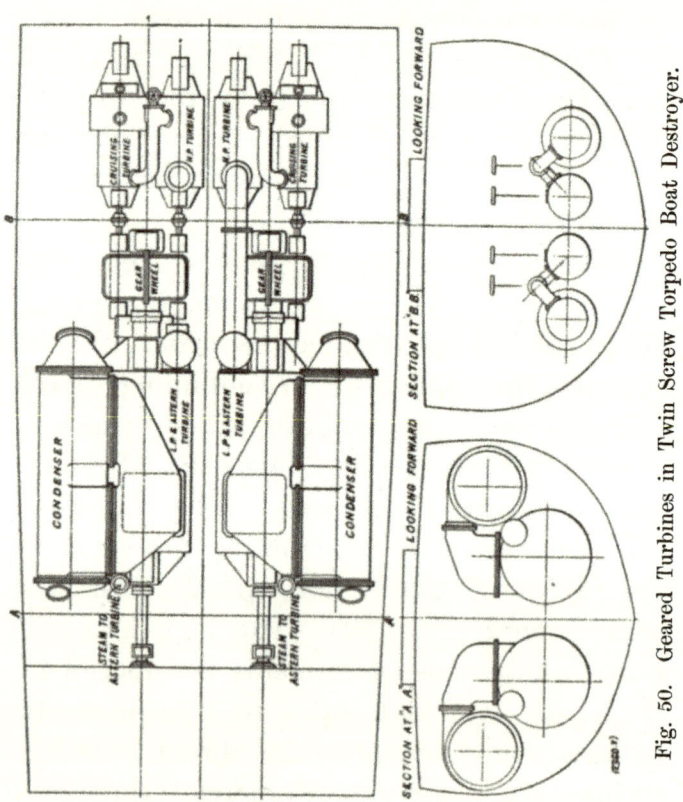

Fig. 50. Geared Turbines in Twin Screw Torpedo Boat Destroyer.

Gearing is also applicable to warships of
the largest size (Fig. 51). It is also finding
a place in cross-channel boats, and two such
vessels for the South Western Railway Co.
are being fitted with all geared turbines like
the "Vespasian." The greatest material
gain, however, will be found in extending
the use of turbines to vessels of slow speed.

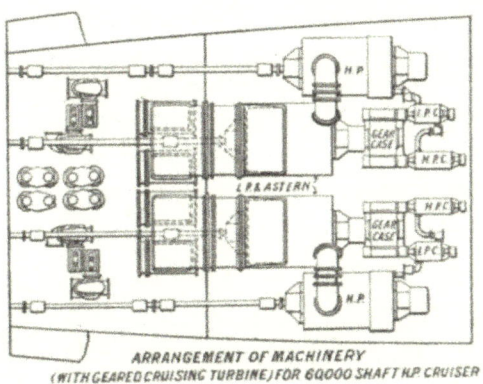

ARRANGEMENT OF MACHINERY
(WITH GEARED CRUISING TURBINE) FOR 60000 SHAFT H.P. CRUISER

Fig. 51.

Half a century ago nearly all screw
vessels had mechanical gearing, one element
being composed of wooden teeth, because the
screw revolved at too high a speed for the

engines. Subsequently it was found practicable to increase the speed of the engines up to that of the screw, and gearing was consequently abandoned. Now very slow speed turbines are found to be incompatible with efficiency, and probably will always be so; accurately cut steel gearing comes to the rescue, and I think will be a permanent institution as long as steam is used to propel our ships.

When we pass through the colliery or iron districts we often see clouds of steam blowing off to waste, but there is much less than was formerly the case, because low pressure turbines worked by the exhaust steam from other engines are coming into extended use for utilizing what was formerly a waste product. They are generally employed for the generation of electricity, or for working blast furnace blowers and centrifugal pumps and gas forcers, but recently an exhaust turbine of 750 horse-power has been applied to driving an iron plate mill in Scotland. It is especially interesting because it is the first turbine to be geared to a rolling mill. The turbine revolves at 2000 revolutions per minute, and by a double reduction of helical gears drives the mill at 70 revolutions. On the same shaft as the rolls is a flywheel of 100 tons weight which helps to equalize the speed. During the short time of each rolling the turbine and flywheel collectively exert 4000 horse-power, the maximum deceleration at the end of each roll being only 7 per cent.

So satisfactory has gearing proved up to the present that it seems probable that by its use turbines will be more widely adopted in the future for many purposes.

In conclusion, I would venture to predict that the use of the land and marine turbine will steadily increase, and that the improvements that are being made to still further increase its economy will for a long time enable the turbine to maintain its present leading position as a prime mover.

For EU product safety concerns, contact us at Calle de José Abascal, 56–1°,
28003 Madrid, Spain or eugpsr@cambridge.org.

www.ingramcontent.com/pod-product-compliance
Ingram Content Group UK Ltd.
Pitfield, Milton Keynes, MK11 3LW, UK
UKHW010851090126
466816UK00011B/159